149322

BISON FARMS

FUNKY FARMS

Lynn M. Stone

The Rourke Corporation, Inc.
Vero Beach, Florida 32964

PHOTO CREDITS:
© Lynn M. Stone: cover, title page, pages 4, 7, 13, 15, 17; courtesy Fermilab: pages 8, 10, 12, 18, 21

ACKNOWLEDGMENTS:
The author thanks Don Hanson and Fermilab of Batavia, Illinois, for its cooperation and support in the preparation of this book.

EDITORIAL SERVICES:
Susan Albury

CREATIVE SERVICES:
East Coast Studios, Merritt Island, Florida

Library of Congress Cataloging-in-Publication Data

Stone, Lynn M.
    Bison farms / by Lynn M. Stone
        p. cm. — (Funky farms)
    Summary: Describes the physical characteristics and habits of American bison and how these large animals are now being raised for their meat on farms across the United States.
    ISBN 0-86593-542-4
    1. Bison farming Juvenile literature. 2. American bison Juvenile literature.
[1. Bison farming. 2. Bison.] I. Title. II. Series: Stone, Lynn M. Funky farms.
SF401.A45S76 1999
636.2'92—dc21   99-25304
                                                                      CIP

CONTENTS

BISON

The bison is one of the most famous and important of American animals. Plains Indians depended on bison meat for food. Bison skins became Native American clothes and homes.

The bison lived largely on grasses. They moved from place to place, always on the lookout for fresh pasture. Some of the wild bison, known as wood bison, lived in more forested lands.

As white settlers moved westward, they **slaughtered** (SLAW turd) millions of the great, shaggy beasts for food and sport.

Bison meant food, clothing, shelter, and even fuel for the Plains Indians of western North America.

In North America, bison are also known as buffalo. True buffalo, however, are certain big, cowlike animals of Africa and Asia.

One difference between these cousins is the number of ribs. A bison has 14 ribs. A buffalo has 13.

A mother bison, known as a cow, gives birth to one calf early in the summer. A male bison, or bull, is a huge, powerful animal. It has short, sharp horns and a shoulder hump.

A bison is amazingly fast for its size. It can run about 40 miles per hour (65 kilometers per hour).

Bison farmers keep a watchful eye on their big bulls. A bull may weigh more than one ton (900 kilograms), and he can double the speed of an Olympic sprinter.

BISON FARMS

Most of the more than 250,000 bison in the United States and Canada live on private farms and ranches. Some of the farmers have "hobby farms." These farmers own just a few bison. Other farmers raise hundreds or even thousands of bison.

Most bison are raised for their meat. Bison tastes much like **beef** (BEEF), the meat of cattle.

In 1972, just four big bison ranches produced most of the bison meat in North America. Now there are hundreds of bison farmers.

This herd of American bison roams in a pasture. Bison calves turn darker as they age.

HOW BISON FARMS BEGAN

The demand for bison meat has made the number of bison farms grow quickly. But the first bison farms were started just to save the bison from disappearing!

Hunters, railroad workers, pioneers, and soldiers killed millions of bison in the 1800s. Suddenly, by the 1880s, the bison were almost **extinct** (ik STINKT)—wiped out completely.

The first bison farmers raised bison to protect them. Their efforts helped save the bison.

The first bison farmers saved the animal from extinction. Today's farms raise bison largely for meat. Note the double fence around this pasture.

Two bison cows rush through the gate at an auction. These cows will be sold to other bison farms where they will produce calves.

Bison or buffalo? This is an African buffalo.
The decoration on its neck is an oxpecker, looking for insects.

WHERE BISON FARMS ARE

Bison farms are scattered throughout North America. Most of the big farms are in western states and in western Canada.

The great herds of wild bison used to live in the West. The two largest wild herds today live in Yellowstone National Park in Wyoming and in Wood Buffalo National Park in Canada's Northwest Territories. Many other herds are kept on huge grassland areas that are fenced.

Most bison herds live on the grassy ranchlands of the West. Here a cow (right) *walks away from a bellowing bull.*

RAISING BISON

A bull bison can weigh 2,500 pounds (1,135 kilograms). That big body needs plenty of food.

Bison farmers graze their herds in large, grassy pastures. Like cattle, bison nibble grass as they roam their pastures. Farmers also feed bison grains and other dry food at certain times of the year.

A young bull stands by a trough (TRAWF) full of food pellets.

Bison are much harder to keep in a pasture than their cattle cousins are. Bison like to wander, and they are very strong. The kind of thin barbed-wire fence used for cattle is like a spaghetti string to bison. Bison farmers usually use metal or wood fences at least six feet (2 meters) tall. Sometimes they use two fences! Bison, remember, are basically wild animals.

Bison cows are raised to produce calves and keep bison herds growing. Bulls are raised largely for meat.

Most bulls are **butchered** (BUHT churd)—killed and turned into meat—at two years of age. By then the animal weighs about 1,200 pounds (545 kilograms).

A bison cow helps her calf to its feet. Calves are born in spring, rarely during snowstorms. Bison are rugged, and snow is no problem.

WHY BISON?

Farmers raise bison mostly because their meat sells well. Bison meat is tasty and it's **lean** (LEEN). Lean meat has little fat in it. In fact, bison meat contains less fat than the meat of pigs, cattle, or even chickens!

The demand for low-fat meat has increased in recent years. That has helped make bison meat more popular and bison farmers more money.

Along with meat, bison farmers sell the bison's **hide** (HIDE), or skin, for leather goods, such as jackets.

Farmers gather at a Midwest bison auction. Live bison are being sold to the farmers who offer the most money for them.

BISON FARMS IN THE FUTURE

The supply of bison meat does not always keep up with the number of customers. That means the number of bison farms is likely to grow in North America.

Bison meat will not replace beef, but it does give meat-eaters another choice.

Bison farming is quite new. Much more needs to be learned about the best ways to raise bison. Many people are studying ways to make bison farming better.

GLOSSARY

beef (BEEF) — the meat of cattle and other cowlike animals

butchered (BUHT churd) — to have been killed, or slaughtered, for meat and other human uses

extinct (ik STINKT) — referring to a kind of plant or animal completely gone from the earth

hide (HIDE) — the skin of an animal

lean (LEEN) — referring to meat with very little fat content

slaughtered (SLAW turd) — having been killed for meat and other human uses; butchered

INDEX

FURTHER READING

Find out more about bison with these helpful books and information sites:

Stone, Lynn, *The Prairie.* Rourke, 1996

Stone, Lynn. *Back from the Edge: The American Bison.* Rourke, 1991

National Bison Association on line at info@nbabison.org and www.nbabison.org